Science Experiments

TIME

by
John Farndon

BENCHMARK **B**OOKS

MARSHALL CAVENDISH
NEW YORK

Marshall Cavendish Corporation

99 White Plains Road

Tarrytown, New York 10591

© Marshall Cavendish Corporation, 2003

Created by Brown Partworks Limited

Library of Congress Cataloging-in-Publication Data

Farndon, John

    Time / by John Farndon
    v. cm. — (Science experiments)
    Includes index.
    Contents: What is time? — Dividing the day — Solar time — Sun
clock — Clocks — Beating time — Calendars — Seasons — Months —
Standard time — Time zones and the date line — Life times — Time and
space — Experiments in science.
    ISBN 0-7614-1470-3
    1. Time measurements—Juvenile literature. 2. Time measurements—
Experiments—Juvenile literature. [1. Time measurements—Experiments.
2. Experiments.] I. Title. II. Series.

QB209.5 .F37 2003
529'.7'078—dc21

                       2002004846

Printed in Hong Kong

Contents

WHAT IS TIME?

Time can be measured with great accuracy. In a marathon, clocks can time exactly how long runners take to finish the race, to within less than a second.

IAAF **ADT** World Marathon Cup
SEIKO
3.18.19
UNISYS COMPUTER
London **ADT** 1991

Did you know?

In Europe, archeologists have found 20,000-year-old bones scratched with lines. The lines are thought to have been scratched by hunters to count the days until the full moon, when it was best to go hunting.

THE HISTORY OF TIME

Space is so vast that light takes a long time to reach us—over two years from the nearest star. When astronomers look at distant stars they are seeing them as they were when their light left them thousands or even millions of years ago. An astronomer's telescope is like a time machine, looking into the past. Light left the most distant detectable galaxies over 13 billion years ago. This is why astronomers believe the universe, and so time itself, is about 14 billion years old.

Looking into space is looking at the history of time.

Our lives are ruled by time. People get up, children go to school, shops open, and many other things happen at certain times of day. Farmers sow their seeds and harvest their crops at particular times of year. Christmas, birthdays, and other festivals are celebrated at the same time each year.

Everyone knows what is meant when someone says time is passing. They know, too, what is meant when someone asks what time it is. Yet it is hard to say exactly what time is.

Long ago, people saw time passing in the way natural things change—how day turns to night, how seasons turn, how plants and animals grow. Now we can see time passing as hands move or figures change on a clock. Modern atomic clocks can measure time passing with astonishing accuracy. Yet scientists and philosophers still find it surprisingly hard to agree on just what time is.

Scientists say time is a dimension (like length and width) and that we move through time just as we can move sideways or backward and forward, or up and down. Time is said to be the fourth dimension; the other three are length, width, and depth.

What makes time different from the other dimensions is that time runs only one way. The clock cannot be turned back. A candle cannot be unburned. We are born, then live, then die—always in that order. This movement of time in one direction is often described as the arrow of time.

DIVIDING THE DAY

The day has always been the most obvious unit of time. Sunrise and sunset are very clear time signals. Our bodies are used to going to sleep at night and waking up in the morning to start a new day. So the day is a unit of time in every culture in the world.

Long ago, people realized that the period of time from midday one day to midday the next is always exactly the same. Time before noon is said to be A.M. or *ante meridiem*, which means before midday in Latin. Time after noon is said to be P.M. or *post meridiem* which means after midday in Latin.

Clocks make it possible to divide the day into smaller units. First, the day can be

Separating the times taken in competitive sports such as rowing needs stopwatches that can measure time accurately in tiny fractions of a second.

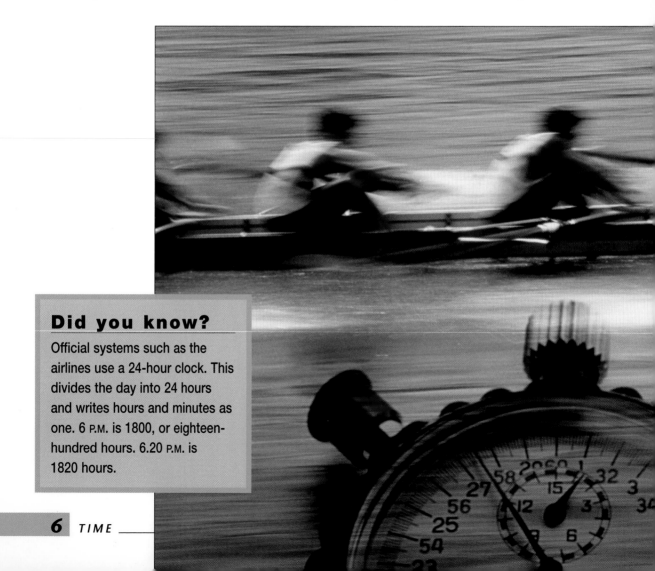

Did you know?

Official systems such as the airlines use a 24-hour clock. This divides the day into 24 hours and writes hours and minutes as one. 6 P.M. is 1800, or eighteen-hundred hours. 6.20 P.M. is 1820 hours.

divided into 24 hours—12 hours between midnight and noon and 12 between noon and midnight.

An hour is divided into 60 minutes, and there are 1,440 minutes in a day. A minute is divided into 60 seconds. Scientists regard the second as the basic unit of time, and measure all other units by it.

Since there are 86,400 seconds in a day, a second was once defined as 1/86,400 of an average day. Now it is defined more accurately using atomic

clocks as 9,192,631,770 vibrations of a cesium-133 atom.

Using modern clocks, even seconds can be divided into smaller units of time. The blink of an eye lasts a *deci*-second, or tenth of a second. A camera shot fast enough to freeze an athlete in action lasts a *milli*-second (a thousandth of a second). A typical home computer takes two *nano*-seconds (billionths of a second) to carry out a basic command. The shortest time that can be measured is a *pico*-second (a trillionth of a second).

SOLAR TIME

The Sun has been humanity's main clock since the very earliest times. Sunrise and sunset dramatically mark the beginning and end of each day. The movement of the Sun through the sky from east to west shows clearly, too, just how far the day has progressed.

It is dangerous to look at the Sun, but it is easy to see how the Sun is moving through the sky

The sundial is the oldest form of timepiece– and among the most reliably accurate!

from the direction of the shadows it casts. It looks as if it is the Sun that is moving, but it is the earth that moves.

As the earth spins around on its axis, it turns us once toward the Sun, then once away again giving us night and day. So the movement of the Sun through the sky during the day is due to the earth's rotation. We are never aware of the earth's movement because the rotation is very steady and everything moves with it, even the air.

Earth always spins the same way, turning eastward. Looking down on the North Pole, the earth turns counterclockwise. This easterly rotation means the Sun always comes up in the east and goes down in the west.

The earth takes exactly 24 hours to turn all the way around. This means the Sun shines from the same direction in the sky once every 24 hours. The time it rises and sets varies through the year, but it always shines from the same direction at the same time of day. Every 24 hours, for instance, it is midday. At this time, the Sun shines from due south in the northern hemisphere and due north in the southern.

As the earth turns, the Sun appears to move through the sky during the day. As it moves, the direction of the shadows it casts changes, moving from west around to east. Sundials use these moving shadows to indicate the time of day.

In focus

SUN AND STAR TIME

Using the direction of the Sun to indicate when the earth has turned once gives the standard, or solar, day of 24 hours. Scientists say a solar day is the time from midday one day to midday the next. But another way to tell how long it takes the earth to turn is to measure how long it takes for stars to return to the same position in the sky each night. Surprisingly this star, or sidereal, day is nearly four minutes less than 24 hours—23 hours 56 minutes and 4.09 seconds. The solar day is longer because Earth does not simply spin on its axis; it travels a little way around the Sun each day. So the earth must rotate an extra 1° before the Sun comes back to the same place in the sky.

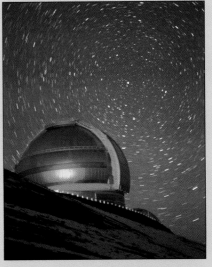

A long camera exposure shows how the stars wheel through the night sky as the earth turns.

SUN CLOCK

You will need

- ✓ A square of thick cardboard (or polystyrene tile)
- ✓ A ruler
- ✓ A triangle
- ✓ A protractor
- ✓ A pencil
- ✓ A black marker pen
- ✓ A craft knife

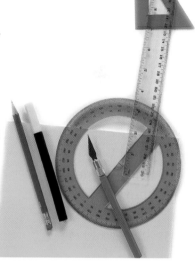

1 Draw a circle on the card. Using the protractor, mark off every 15° either side of the 90° vertical—8 marks each side.

In focus

THE HOUR ANGLE

The 15° for each hour in the experiment is only rough. For accuracy, a sundial must be adjusted for the latitude where you are in the world. To work out the correct angle for each hour, use the following equation with school math tangent (*tan*) and sine (*sin*) tables. A is the angle of the line, L is your latitude, and H is the hour times 15. The equation is $\tan A = \sin L \times \tan H$.

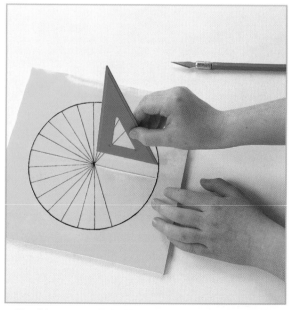

3 Use a craft knife to cut a slot along the 90° vertical line. Then carefully push in the triangle so it stands upright.

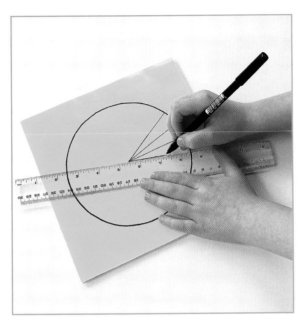

2 With a ruler, draw a line from the center of the circle to each of the 15° marks. Each line marks an hour of the day.

What is happening?

Shadows show clearly how the Sun moves, and the sundial uses them to show the time of day. It only works on sunny days, of course, when there is a strong, clear shadow. The time is indicated by the place where the shadow of the upright, called the gnomon, falls on the dial. Although the Sun rises and falls at different times during the year, it always shines from the same direction at the same time of day. So if the shadow from the Sun falls on your eleven o'clock mark you can be sure it is eleven o'clock whatever time of year—providing the board is set up in the same direction, facing the midday Sun.

Set the completed sundial on a sunny windowsill or out in the open on a calm, sunny day. Place it so that the triangle side always faces the midday Sun, and the shadow of the triangle is a thin line falling exactly along the center line. The movement of the Sun's shadow around the board tells you the time of day. You can see if the marks are right by checking them on a clock as the Sun moves around.

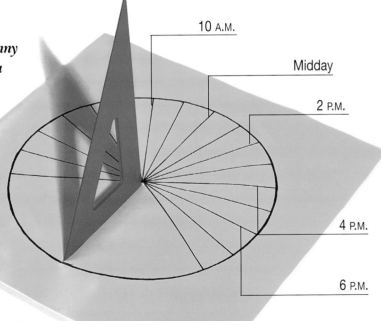

10 A.M.

Midday

2 P.M.

4 P.M.

6 P.M.

CLOCKS

The Great Clock of Big Ben in London's Houses of Parliament is one of the world's most famous clocks. Coins added every now and then to the timing weight have helped it keep almost perfect time ever since it was installed in 1854.

Did you know?

Before modern satellite systems, ships needed a very accurate clock or chronometer to work out their longitude at sea. The first chronometer was made by English clockmaker John Harrison in 1761.

To keep time, clocks need a timer. This can be a steady flow, a steady change, or a steady beat. Early water clocks relied on a steady flow or drip of water. Ancient sandglasses, like kitchen eggtimers, relied on a steady flow of fine sand. But no early timer was accurate enough to measure seconds.

The breakthrough came in the 1600s when it was realized that a swinging pendulum could make a very accurate timer. The invention of the pendulum clock by Dutch scientist Christiaan Huygens in 1656 created a revolution in timekeeping.

With Huygens's clock, seconds could be measured for the first time. The clock also kept time to within a minute a day. So for the first time in history, people could tell the time exactly, all day and night, whatever the weather. Soon lives were ruled by clocks, as shops opened, people started work, and did many other things according to the clock.

By the 1890s, the best pendulum clocks could keep time to within a hundredth of a second a day. Even greater timekeeping accuracy was developed with the use of electricity as a power supply. Electrical outlets not only supply power, but the regular beat of the electricity supply also regulates a clock's speed.

In focus

HOW PENDULUM CLOCKS WORK

In a pendulum clock, the hands are driven by a falling weight. The weight hangs from a cord wrapped around a drum that is pulled slowly around as the weight falls. The drum's rotation turns the hands at different rates via a series of gears. Once the weight has fallen all the way, the clock stops until the weight is wound up again. The key to the clock's accuracy is the pendulum. A pendulum is a weight that swings to and fro on the end of the line at a very steady pace. As it swings, it releases a brake on the falling weight called an escapement. On each swing, the escapement releases the weight, letting it fall briefly before stopping it again. The swinging pendulum thus ensures that the weigh falls slowly and steadily, keeping the clock in time.

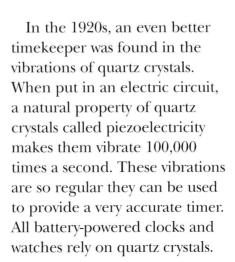

Escapement rocked by swinging pendulum

The rocking escapement tooth stops the wheel turning and lets it go alternately

Wheel turned by falling weight

Swinging pendulum

In the 1920s, an even better timekeeper was found in the vibrations of quartz crystals. When put in an electric circuit, a natural property of quartz crystals called piezoelectricity makes them vibrate 100,000 times a second. These vibrations are so regular they can be used to provide a very accurate timer. All battery-powered clocks and watches rely on quartz crystals.

BEATING TIME

You will need

- ✓ A plastic soda bottle
- ✓ Scissors
- ✓ A black marker pen
- ✓ A clock that shows seconds
- ✓ A drawing compass
- ✓ Water

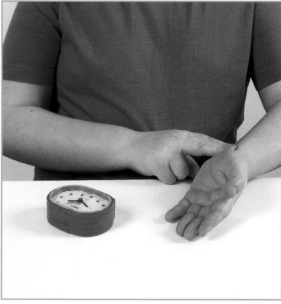

1 Take your pulse by laying two fingertips on the artery on the inside of your wrist. Time the pulses with a clock.

In the real world

GALILEO'S BELL-ROPE

Until recently, the regular beat used for most good clocks was the swing of a pendulum. The first person to realize this was the great Italian scientist Galileo in the 1590s. Galileo was watching a swinging bellrope in church in Pisa one day. Timing each swing against his own pulse, he realized that each swing was perfectly regular. He found this was true whatever the size of the swing. When the swinging rope slowed down, it simply swung less far and so took the same time to complete a swing. This is why pendulums keep such good time.

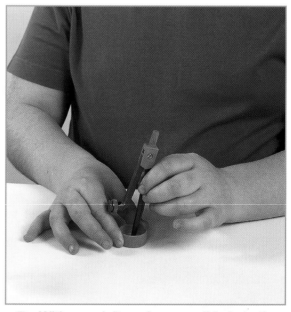

3 With an adult, make a small hole in the lid of the bottle with a compass and screw the lid on to the bottle.

What is happening?

All clocks need a regular beat or a steady flow to keep them in time. Your pulse is a simple natural beat. Pulses vary, but if you are standing still, each pulse should last a little more than a second. By counting pulses, you can time how long something lasts in seconds. The drips from the water clock can be used in the same way once you have timed how frequent they are. The level of water in the bottle provides a measure of a longer period of time. Moreover, it provides a record of time passing, so you can leave the clock, come back and see how much time has passed.

2 Now start to make a water clock, by cutting the soda bottle in half with a pair of scissors.

Turn the top half of the bottle upside down and fill with water. If necessary, enlarge the hole in the lid until water just begins to drip through. Rest the bottle top inside the bottom half of the bottle to catch the water. Using a clock, count how many drips fall in a minute. Then mark the level of water in the bottle at 5 minute intervals. You have now made a simple water clock.

CALENDARS

ach culture has its own calendar and calendar year. Muslims, for instance, base their calendar mainly on the cycles of the moon—between 30 and 29 days. So the Muslim year of 12 months lasts either 354 or 355 days. The Hindu year last 360 days. The Jewish year varies between 353 and 385 days.

The western calendar used over much of the world depends on the annual journey of the earth around the Sun. Because Earth takes just over 365 days to travel around the

The mysterious monument of Stonehenge in England is at least 4,000 years old. No one knows what it was for, but its stones align perfectly with the movements of the Sun which mark the seasons.

Did you know?

More than 5,000 years ago, the Mayans of Central America established an accurate 365-day calendar based on the movements of the Sun, the Moon, and the planet Venus.

Sun, the western calendar year is 365 days. This means the Sun is always in much the same place in the sky on the same date each year.

However, the calendar and Sun are not quite in step. The earth does not take exactly 365 days to circle the Sun; it takes 365.252199 days. Because the earth wobbles very slightly, its orbit varies by a few seconds each year. So it is very difficult

In the real world

Our calendar originated with the Roman emperor Julius Caesar.

CALENDAR MAKERS

The idea of a 365-day year with leap years every fourth year was introduced by the Roman emperor Julius Caesar in 46 B.C. The Julian calendar provides the basis for our calendar today. By 1582, the 11 "extra" minutes not balanced by the leap year meant the Sun reached the equinox (the first day of spring) ten days earlier than the calendar. To make up for this, Pope Gregory introduced the century year and moved the calendar forward 10 days, going straight from October 4 to October 15. Protestants, already in rebellion against the Pope, rioted because they thought Gregory had cut ten days from their lives.

for scientists to work out exactly how long a year lasts.

Compensating for these slight differences has always been a problem for calendar makers. The extra 0.25 days is made up by simply adding an extra day every fourth, or leap, year. But the Sun then arrives back in the same place about 11 minutes earlier each year. To make up for this, one leap year day is cut every 400 years. This is called a century year. Even so, the calendar remains a few seconds off each year.

SEASONS

You will need

- ✔ Colored adhesive tape
- ✔ A large ball
- ✔ Scissors
- ✔ A flashlight
- ✔ A bowl (to support the ball)

1 Use the ball's seam to help you stick tape exactly around the middle to be an Equator, or draw a line with a marker.

In focus

SOLSTICES AND EQUINOXES

As Earth circles the Sun during the year, the Sun appears to rise and fall in the sky as each place on Earth turns to face it more or less directly. The Sun reaches its zenith, or highest point, on about June 22 in the northern hemisphere. This is the summer solstice, the longest day of the year. It sinks lowest on December 22. This is the winter solstice, the shortest day. Midway between are the vernal (spring) and autumnal (fall) equinoxes, when the day and night are both equally long (12 hours) all over the world.

3 Set the flashlight on a tin or book at the same height as the ball. Turn the ball so it shines on the Equator line as shown.

What is happening?

Here, you move the ball around the torch as Earth moves around the Sun during the year. The tilting Equator line shows why Earth's tilt gives us seasons. Only in spring and fall (Step 3), when the earth is on opposite sides of the Sun, does the Sun strike the Equator directly. In between, the Sun strikes Earth directly to the south of the Equator in December (main shot), making the southern hemisphere warmer, but bringing winter to the north. In June, the Sun strikes directly to the north of the Equator, bringing long, warm summer days to the northern hemisphere, while the southern hemisphere has its winter.

2 Rest the ball in a bowl so the line tilts, like the Equator on a globe. The angle should 23°, but you need not be exact.

Move the ball in a circle around the flashlight. Take care to keep the ball facing the same direction, rather than turning it. Stop at each quarter, and rotate the flashlight to face the ball. At each quarter the flashlight shines at a different place relative to the ball's Equator.

MONTHS

You will need

- ✓ A white ball to represent the Moon
- ✓ A flashlight to represent the Sun

1 Hold the Moon ball out to your side at arm's length. Ask a friend to shine a flashlight on the ball from in front of you.

What's happening

The Moon has no light of its own but shines because one side is caught in sunlight. In the experiment, you move the Moon ball around you to mimic the way the Moon circles Earth—and see how it seems to change shape as we see different amounts of its sunlit side. Step 1 shows a half moon, when we see half the sunlit side. Step 2 is an old moon when we see just a thin crescent. Move the ball all the way around you to see the complete sequence from crescent-shaped new moon to the full moon, and back to the old moon. This sequence takes about 29 days, and was the origin of our months or "moonths."

2 Swing your arm around in front of you. Only a sliver of the ball is lit. Try circling the ball around you at arm's length.

3 A third of the way around, the same side of the Moon still faces Earth, yet it has turned a third on its axis.

Like Earth, the Moon spins around, but takes 27.3 days to turn once. This is exactly the same time it takes to circle the earth. It is also why we always see the Moon from the same side, as this experiment shows. Although the Moon takes 27.3 days to circle Earth, it takes 29 days to go from one full moon to the next, because Earth moves around the Sun a little, slightly delaying the phases.

4 Half way around, the same side of the Moon still faces the earth, yet it has turned half on its axis.

2 A sixth of the way around, the same side of the Moon faces Earth, yet it has turned a sixth on its axis, as you can see by looking at the rest of the room.

5 Two thirds of the way around, the same side of the Moon is still facing Earth but has turned two-thirds on its axis.

1 As you start, note which side of the Moon ball is facing the earth.

Ask a friend to stand in the middle to represent the earth, while you hold the Moon ball in front of you. Slowly sidestep around your friend, staying facing her or him all the time. See how as you go around, you also turn to face every direction in the room.

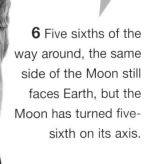

6 Five sixths of the way around, the same side of the Moon still faces Earth, but the Moon has turned five-sixth on its axis.

STANDARD TIME

The Greenwich Meridian provided the standard time for all the world from 1884 to 1972.

As the earth spins around, the Sun is always coming up on a new part of the world—and setting halfway around the world. When it is dawn in New York, it is sunset the other side of the world in China. It is also midday in Europe at this time, and midnight in the middle of the Pacific Ocean.

Two hundred years ago, every place in the world set its own time system, according to when the Sun came up locally. Even countries as small as Great Britain had dozens of local time systems. As people began to travel the world more and more, this began to cause confusion. So in 1884, the countries of the world agreed to adopt the same standard time, Greenwich Mean Time (GMT), based on the time kept at the Royal Greenwich Observatory in London, England.

To make it easier to set clocks, the world was split into 24 time zones based on GMT. One zone for each hour of the day. As people travel east around the world, they put their clocks forward one hour for

ATOMIC CLOCKS

Just as a drum vibrates at a particular pitch or frequency, so do atoms and molecules. These vibrations are so regular that they can be exploited to make the world's most accurate clocks, atomic clocks. These special clocks, all housed in special laboratories, mostly use the cesium-133 atom. Since 1967, a second has been defined as 9,192,631,700 vibrations of a cesium-133 atom.

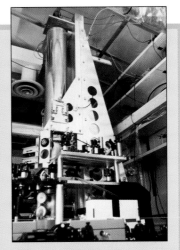

Atomic clocks keep time to within one millionth of a second per year.

each zone they pass through—until they reach a line running through the Pacific Ocean called the international date line. When people cross the Date Line, they continue to put the clock forward, but put the calendar date back by one day.

Each time zone is about 15° of longitude. But the borders are bent here and there to match state and national boundaries, so they won't fall in awkward places. GMT is the time at 0° longitude. The international date line roughly follows the 180° line. The USA is so big that it stretches a third of the way around the world, through eight time zones—Atlantic, Eastern, Central, Mountain, Pacific, Alaska, Hawaii-Aleutian, and Samoa.

Improving accuracy of clocks in the 20th century meant that

GMT was no longer adequate. So in 1972, countries agreed to replace it with Coordinated Universal Time (UTC). UTC depends on atomic clocks kept in laboratories around the world, run mostly by the United State's National Institute of Standards and Technology (NIST) agency. NIST sets the Official United States Time. Most western countries move the clock forward one hour in summer. In the U.S.A, this is called Daylight Saving Time.

Did you know?

NIST's atomic clocks now depend on atoms vibrating nine billion times a second. Soon they will use light vibrating 100,000 times faster—at one quadrillion times a second!

TIME ZONES AND THE DATE LINE

You will need

- ✔ A globe
- ✔ A flashlight
- ✔ Colored tape

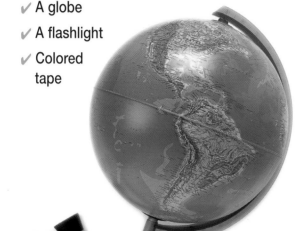

1 In a dark room, ask a friend to point the flashlight at the globe. Move around the globe until you just see the flashlight.

In the real world

THE INTERNATIONAL DATE LINE

The international date line (i.d.l.) is the imaginary line on the earth that separates two consecutive calendar days. The world to the east of this line is always one day ahead of the world west of the line. Without this date line, someone traveling around the world westward would find they arrived a day later than they counted. This happened to the crew of Magellan's ship, the first to sail around the world, in 1523. The i.d.l. could be anywhere on the globe, but countries have agreed that it should be halfway around the world from Greenwich, England (180° latitude) in the Pacific Ocean.

All the world has the same day only when it is midnight at the i.d.l. As the world turns on, some of the world remains the old day, but the rest moves on to a new day, as this diagram shows.

Sun

Earth's rotation

Noon

New day

int. date line

New day

Midnight

Old day

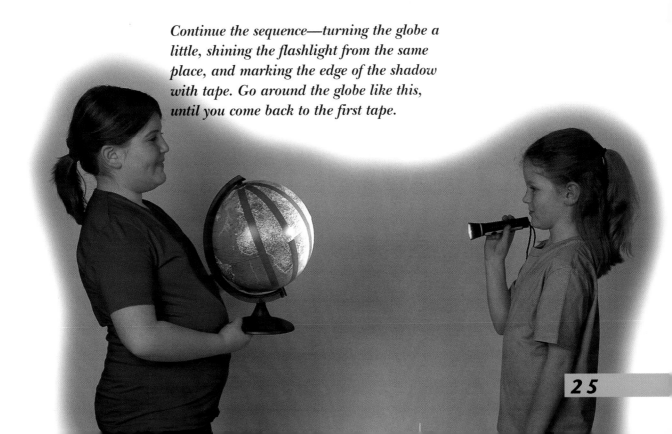

What is happening?

At any time, half the world is sunlit day, and half is in the shadow of night. Wherever the edge of the shadow falls on the globe, it is either sunrise there in the world (at the western edge) or sunset (at the eastern edge). The tape marks the edge of the sunrise. As you turn the globe and mark a new shadow edge, it is like the world turning and the Sun rising in a new place. The Sun rises two hours later for every 30° westward around the world. Once you have marked twelve 2-hour sunrises, you should have turned the globe once around, completing a 24-hour day. You have made your own time zones and date line.

2 Shine the light at the globe. Tape along the edge of the half of the globe in shadow. Turn the globe 30° and repeat.

Continue the sequence—turning the globe a little, shining the flashlight from the same place, and marking the edge of the shadow with tape. Go around the globe like this, until you come back to the first tape.

LIFE TIMES

All living things have their own in-built, biological clocks that control the pattern of their lives and stimulate them to behave in certain ways at certain times.

A creature's biological clock often works on regular cycles, or biorhythms. These fit in with natural cycles such as day and night, the seasons, and the rise and fall of tides. By far the most powerful of biological timers, however, is the clock controlling an animal's life span—telling it when to grow, when to mate, when to age, and when to die.

No animal can escape the ticking clock that begins at birth, and brings their life to an end a certain time later.

Nematode worms are timed to end their lives in just a few weeks. Right whales may live more than two centuries. Research has shown that every creature lives for about one billion heartbeats. So tiny mice whose hearts beat very fast live only a few years, while humans, whose hearts beat much slower, can live 80 or more years.

Humans go to sleep each night as the hypothalamus, the body's internal alarm clock in the brain, sends out sleep signals.

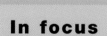

In focus

JET LAG

When air travelers fly across time zones, they can suffer from "jet lag" as their internal biological clock becomes at odds with the new timing of day and night in their destination. The internal clock is set by light falling on the eyes, which tells the brain to either cut or boost body levels of the hormone melatonin which controls sleep patterns. In jet lag, the coming of daylight at the wrong time disrupts the normal rhythm of melatonin release, making people feel tired and disoriented.

After they reach the age of puberty at 12 years old or so, girls and women experience a monthly rhythm called the menstrual cycle. This prepares a new egg each month ready to grow into a baby if fertilised.

The most obvious rhythms are daily, or circadian, and control when animals sleep, wake, eat, and much more. Animals may be active mostly during the day (diurnal) or active mostly at night (nocturnal).

Like all diurnal animals, humans have a biological clock

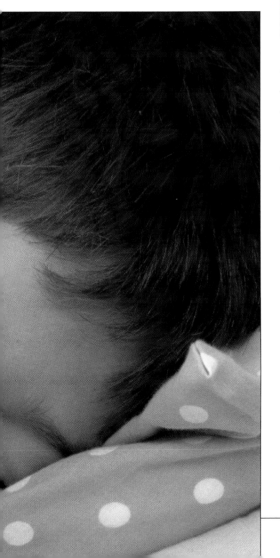

Did you know?

In fall, the leaves on many trees change color and drop off. Trees lose their leaves to cut their need for water in winter when water may be frozen. The timing of the change is controlled mainly by nighttime temperatures so can be delayed by a warm spell.

Few biological rhythms produce more spectacular effects than the red, gold, and brown fall colors of leaves on trees.

in their brain that tells them when to sleep and when to wake. In humans, there is an area called the SCM in the hypothalamus region of the brain. The SCM is a natural clock which works so well that people often wake up just before their alarm clock.

Like many biological clocks, this works through special chemical messengers called hormones that circulate around the body. The hypothalamus makes us sleepy by sending signals to the nearby pineal gland to release the hormone melatonin. Scientists have found that a small injection of melatonin is enough to send people off to sleep.

TIME AND SPACE

For the great 17th century scientist Isaac Newton (1642-1727), time and space were completely immobile backgrounds against which everything in the universe moved. He believed that time is absolute—that is, it keeps ticking away at exactly the same rate everywhere in the universe all the time. For all practical purposes, this idea of absolute time is still right. But a century ago the brilliant scientist Albert Einstein (1879-1955) showed the truth is more complicated with his theories of Relativity.

Einstein showed that time is relative, not absolute—that is, it depends entirely on where you measure it. Light takes millions of years to reach us from distant

If you took a very accurate clock on board a high speed train, you would find it would lose a little time during the journey.

Did you know?

If time travel were possible, it would create all kinds of contradictions, or paradoxes. For instance, if you were to travel back in time before your parents were born and prevent them from getting married, then you would never have been born. But if you didn't exist, who kept your parents from getting married?

stars, so we see them not as they are but as they were millions of years ago. People elsewhere in the universe would see them at a different time. Light takes a little time to reach us even from nearby things, so time depends on where you measure it.

Einstein showed that time cannot really be separated from movement through space. In some ways, time is movement through space, and can be thought of as part of the same thing, which Einstein decided to call spacetime.

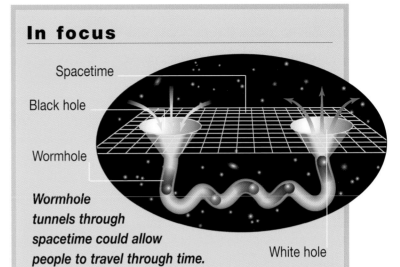

In focus

Spacetime

Black hole

Wormhole

Wormhole tunnels through spacetime could allow people to travel through time.

White hole

TRAVELING THROUGH TIME

Einstein showed that time is just another dimension and runs at different speeds in different places. Ever since, scientists have wondered whether we could travel through time to the past or future. In the 1930s, the American mathematician Kurt Gödel showed it might be done by bending spacetime. Spacetime is bent very powerfully by regions in space called black holes. Black holes have such strong gravity they suck everything in, including light. Some scientists think tiny black holes may be linked to reverse black holes called white holes by a tunnel through spacetime called a wormhole. It might be possible to time-travel by opening up a wormhole and slipping through it.

One effect of this is that the faster things travel, the slower time runs. When *Apollo 11* went to the Moon, an accurate clock on board lost a few seconds, not because of any fault but because time ran slowly in the speeding space ship. When things are moving almost at the speed of light, time slows almost to a standstill.

Experiments in Science

Science is about knowledge: it is concerned with knowing and trying to understand the world around us. The word comes from the Latin word, *scire*, to know.

In the early 17th century, the great English thinker Francis Bacon suggested that the best way to learn about the world was not simply to think about it, but to go out and look for yourself—to make observations and try things out. Ever since then, scientists have tried to approach their work with a mixture of observation and experiment. Scientists insist that an idea or theory must be tested by observation and experiment before it is widely accepted.

All the experiments in this book have been tried before, and the theories behind them are widely accepted. But that is no reason why you should accept them. Once you have done all the experiments in this book, you will know the ideas are true not because we have told you so, but because you have seen for yourself.

All too often in science there is an external factor interfering with the result which the scientist just has not thought of. Sometimes this can make the experiment seem to work when it has not, as well as making it fail. One scientist conducted lots of demonstrations to show that a clever horse called Hans could count things and tap out the answer with his hoof. The horse was indeed clever, but later it was found that rather than counting, he was getting clues from tiny unconscious movements of the scientist's eyebrows.

This is why it is very important when conducting experiments to be as rigorous as you possibly can. The more casual you are, the more "eyebrow factors" you will let in. There will always be some things that you can not control. But the more precise you are, the less these are likely to affect the outcome.

What went wrong?

However careful you are, your experiments may not work. If so, you should try to find out where you went wrong. Then repeat the experiment until you are absolutely sure you are doing everything right. Scientists learn as much, if not more, from experiments that go wrong as those that succeed. In 1929, Alexander Fleming discovered the first antibiotic drug, penicillin, when he noticed that a bacteria culture he was growing for an experiment had gone moldy—and that the mold seemed to kill the bacteria. A poor scientist would probably have thrown the moldy culture away. A good scientist is one who looks for alternative explanations for unexpected results.

Glossary

biorhythm: Regular cycle of change in living things, such as sleeping and waking.

black hole: A point in space with such powerful gravity that it sucks in all energy and matter, making a hole in spacetime.

calendar: System for working out the days and months of the year.

century year: A leap year that is cut every 400 years in order to keep the calendar in time with the movement of the earth around the Sun.

circadian rhythm: Pattern repeated every 24 hours.

date line: imaginary line running from the North Pole to the South Pole dividing one calendar date from the next.

dimension: Measurement in a particular direction such as length, width, or height.

diurnal: Either occurring during the day rather than the night or ocurring once every day.

equinox: One of two days each year when the midday sun is directly overhead at the Equator, making day and night equal length all over the world.

gnomon: The needle of a sundial which shows the time of day by the direction of its shadow.

hemisphere: Half of the earth to the north or south of the Equator.

latitude: An imaginary line around the earth joining all points the same distance from the equator, or more loosely: distance north or south of the equator.

leap year: Year in which an extra day is added to keep the calendar in step with Earth's movement around the Sun.

longitude: An imaginary line around the earth joining all points the same distance east or west of the prime meridian—a line that runs from the North Pole to the South Pole through Greenwich, England.

lunar month: The time between one new moon and the next, about 29.5 days.

meridian: Line of longitude, or old word for midday.

meridiem: Latin for midday.

pendulum: Weight swinging freely from a fixed point.

piezoelectricity: Electric current produced by certain crystals under pressure.

second: 1/60 of a minute or 9,192,631,770 vibrations of a cesium-133 atom.

sidereal: About the stars. A sidereal day is measured by the movement of the stars.

solstice: The longest or shortest day of the year, when the overhead sun is farthest from the Equator.

spacetime: The entire spread of time and space—adding time to the normal three dimensions of space.

time zone: One of 24 zones where time is the same.

Index